W9-AFQ-012
Tomato seeds.
Tomato seeds.

YOUR PLANET NEEDS YOU!

Philip Bunting

BLOOMSBURY
CHILDREN'S BOOKS
NEW YORK LONDON OXFORD NEW DELHI SYDNEY

No waste in the wild.

Let's start at the beginning. In the natural world, there is no waste. Zero. Zip. Zilch. Everything arising naturally from the earth is eventually broken down and reused in a new way.

No part of any plant, plankton, person, or parrot is wasted. All of the bits that make up each living thing are eventually returned to the earth to help make new life.

Here's an example:

1. Distant Ancestor eats banana, carelessly discards peel.
2. Distant Ancestor unexpectedly slips on peel. Dead.
3. Bacteria begin to break down Distant Ancestor's body, slowly returning it to the earth.
4. Fungi, ants, worms, and other decomposers break down the body further, releasing all of its nutrients into the soil.
5. Some time later . . . fertile, nutrient-rich soil makes a lovely home for a new banana seed.
6. The seed sprouts to become a seedling.
7. Seedling ingests the nutrients in the soil to become a handsome new banana tree.
8. Distant Ancestor does not heed the lessons.

Incredibly **serious** disclaimer: No distant ancestors were harmed in the making of this book.

Why is there so much trash in the world today?

For most of human history, almost everything we used, ate, made, or played with came directly from the earth, so we created very little waste.

But a few hundred years ago, we industrious humans began industrializing. For the first time in our history, we began making things with materials that were not found in nature. And we began making them on a really big scale.

The more stuff we make, the more waste we produce. Today we make more stuff than ever before. As a result, we're making heaps more garbage than ever before (in every sense).

Every year, each of us creates more than a ton of waste.
That's about the weight of an adolescent hippo!
Now think about all the people in your family,
your school, your neighborhood, your town . . .
that's a lot of stinky, stinky hippos.
Hey, leave the hippos out of this.

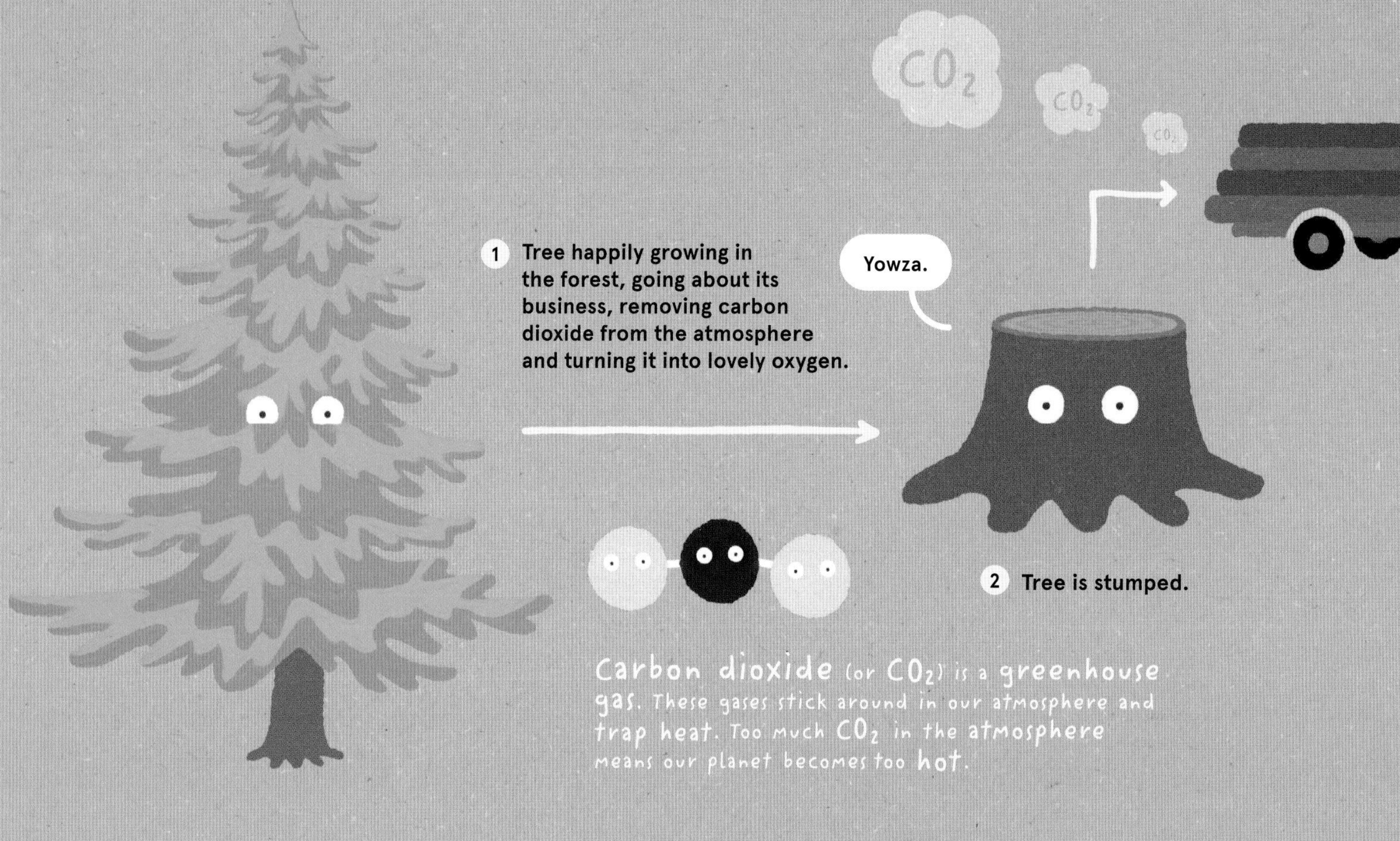

Why is waste so bad for our planet?

Waste is created when we make stuff, and also when we're done with it. Let's take a sketchbook as an example. Waste is produced at almost every stage of the sketchbook's life. Unfortunately, this is true of pretty much everything we use, eat, wear, or play with.

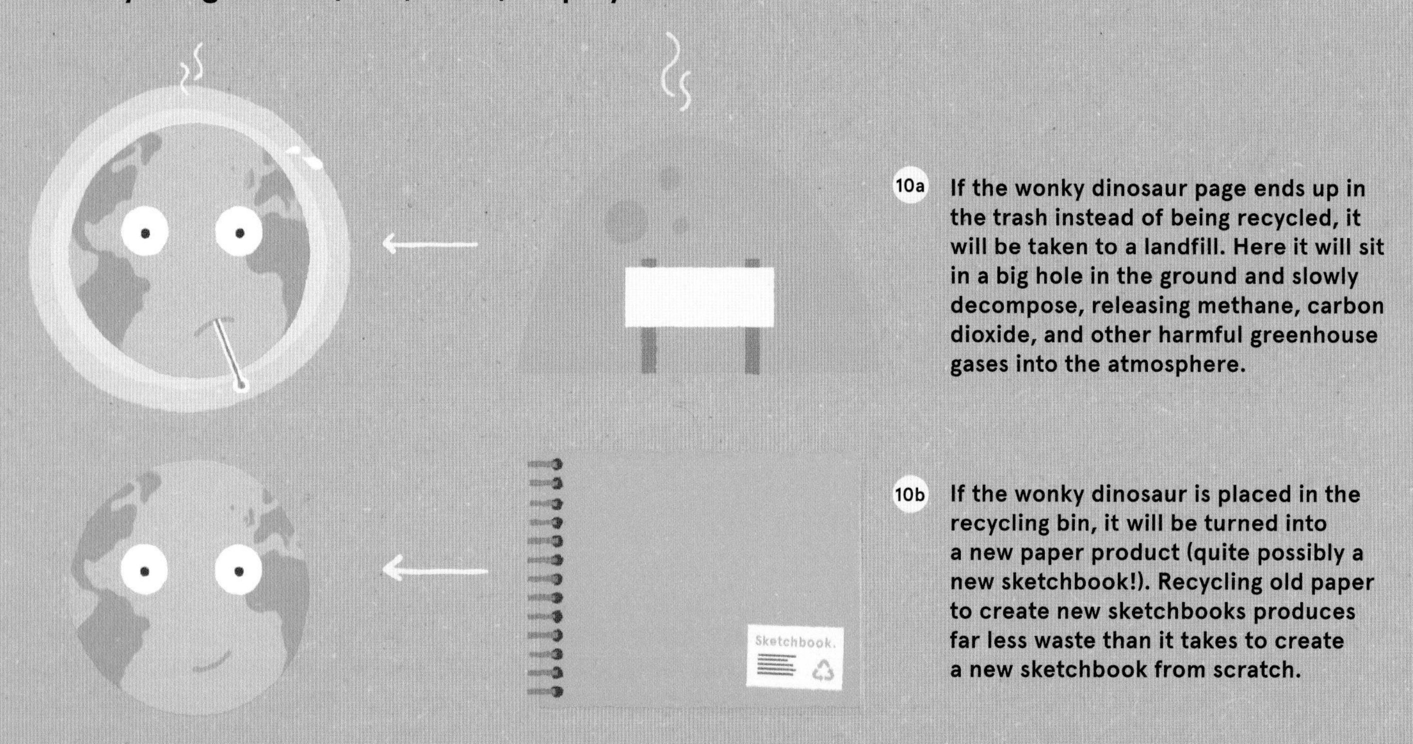

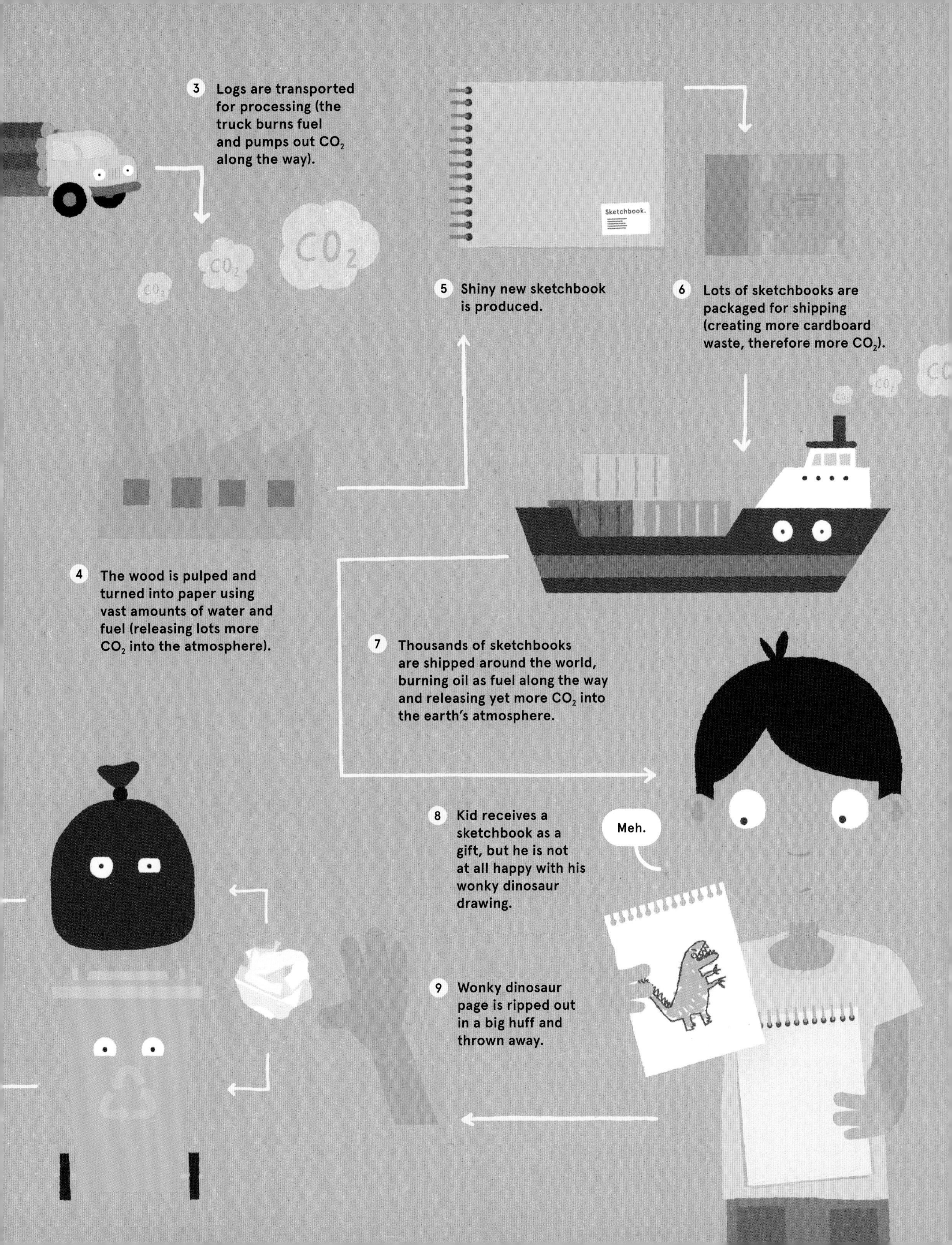

3 Logs are transported for processing (the truck burns fuel and pumps out CO_2 along the way).
CO_2
CO_2
CO_2
Sketchbook.
5 Shiny new sketchbook is produced.
6 Lots of sketchbooks are packaged for shipping (creating more cardboard waste, therefore more CO_2).
4 The wood is pulped and turned into paper using vast amounts of water and fuel (releasing lots more CO_2 into the atmosphere).
7 Thousands of sketchbooks are shipped around the world, burning oil as fuel along the way and releasing yet more CO_2 into the earth's atmosphere.
8 Kid receives a sketchbook as a gift, but he is not at all happy with his wonky dinosaur drawing.
Meh.
9 Wonky dinosaur page is ripped out in a big huff and thrown away.

What kinds of trash do we produce at home?

Waste is an unavoidable part of all of our lives. It is really important for us to understand the waste we produce so we can begin to do something about it.

As with many things in our wonderful world, the best place to begin making a difference is at home. Here are the main kinds of household garbage thrown out by a typical family in an ordinary week:

Paper.

We throw out a lot of paper and cardboard, often in the form of packaging, but also in the form of sketchbooks, printer paper, receipts, and so much more.

Food waste.

This huge pile is made up of food that has passed its sell-by date, as well as uneaten food and offcuts of edible stuff from food preparation.* It is estimated that food waste accounts for up to half of all household waste in developed countries.

*You may note that there is no ice cream in this pile. Very little ice cream seems to get wasted. No idea why.

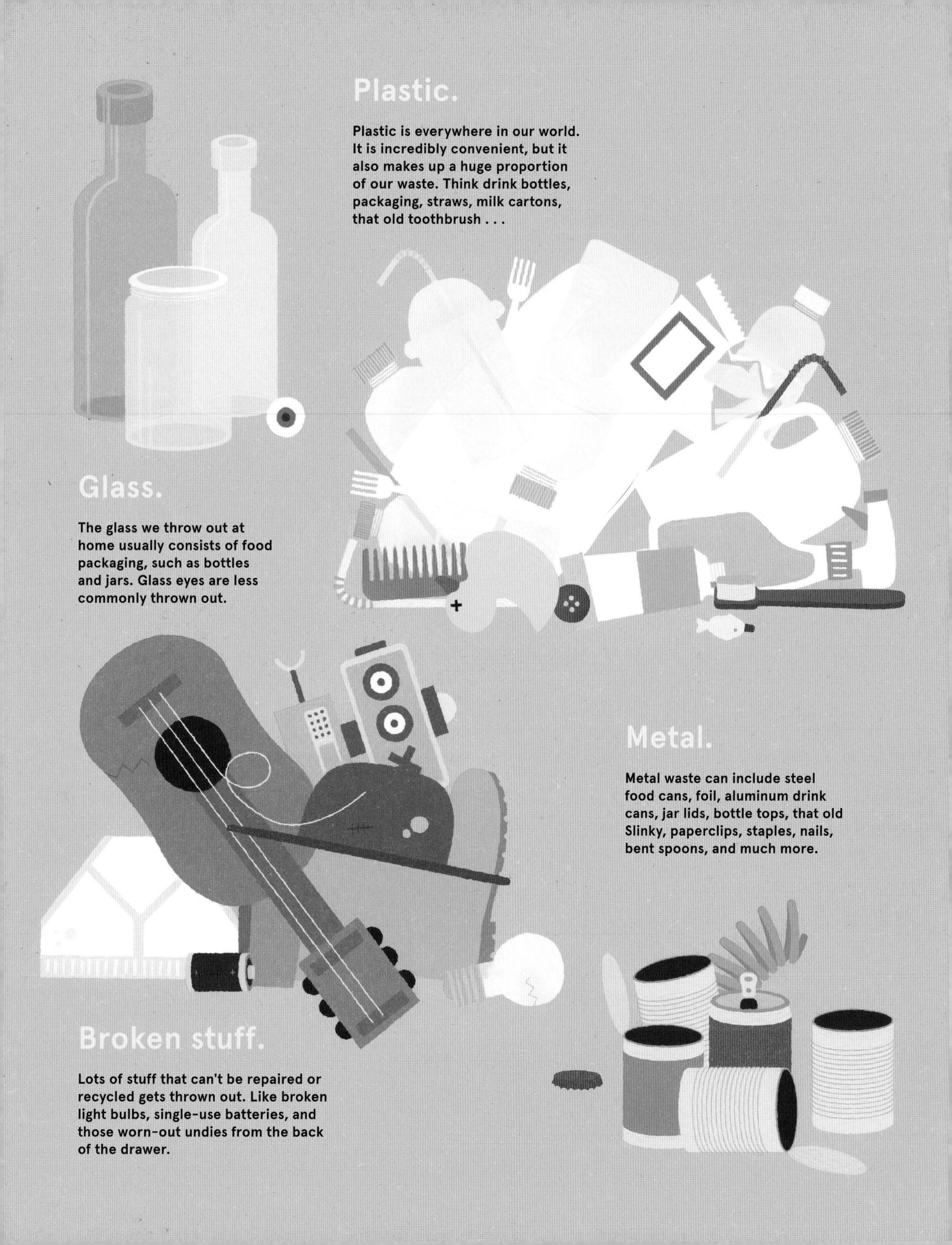

Plastic.

Plastic is everywhere in our world. It is incredibly convenient, but it also makes up a huge proportion of our waste. Think drink bottles, packaging, straws, milk cartons, that old toothbrush . . .

Glass.

The glass we throw out at home usually consists of food packaging, such as bottles and jars. Glass eyes are less commonly thrown out.

Metal.

Metal waste can include steel food cans, foil, aluminum drink cans, jar lids, bottle tops, that old Slinky, paperclips, staples, nails, bent spoons, and much more.

Broken stuff.

Lots of stuff that can't be repaired or recycled gets thrown out. Like broken light bulbs, single-use batteries, and those worn-out undies from the back of the drawer.

How long does our trash take to decompose?

At present, most of our household garbage is sent to a landfill (a big hole in the ground). In the landfill, our junk slowly begins to break down into smaller and smaller pieces until it no longer looks like the thing we chucked in the trash. Depending on what the junk is made from, it can hang around for a really, really, really long time.

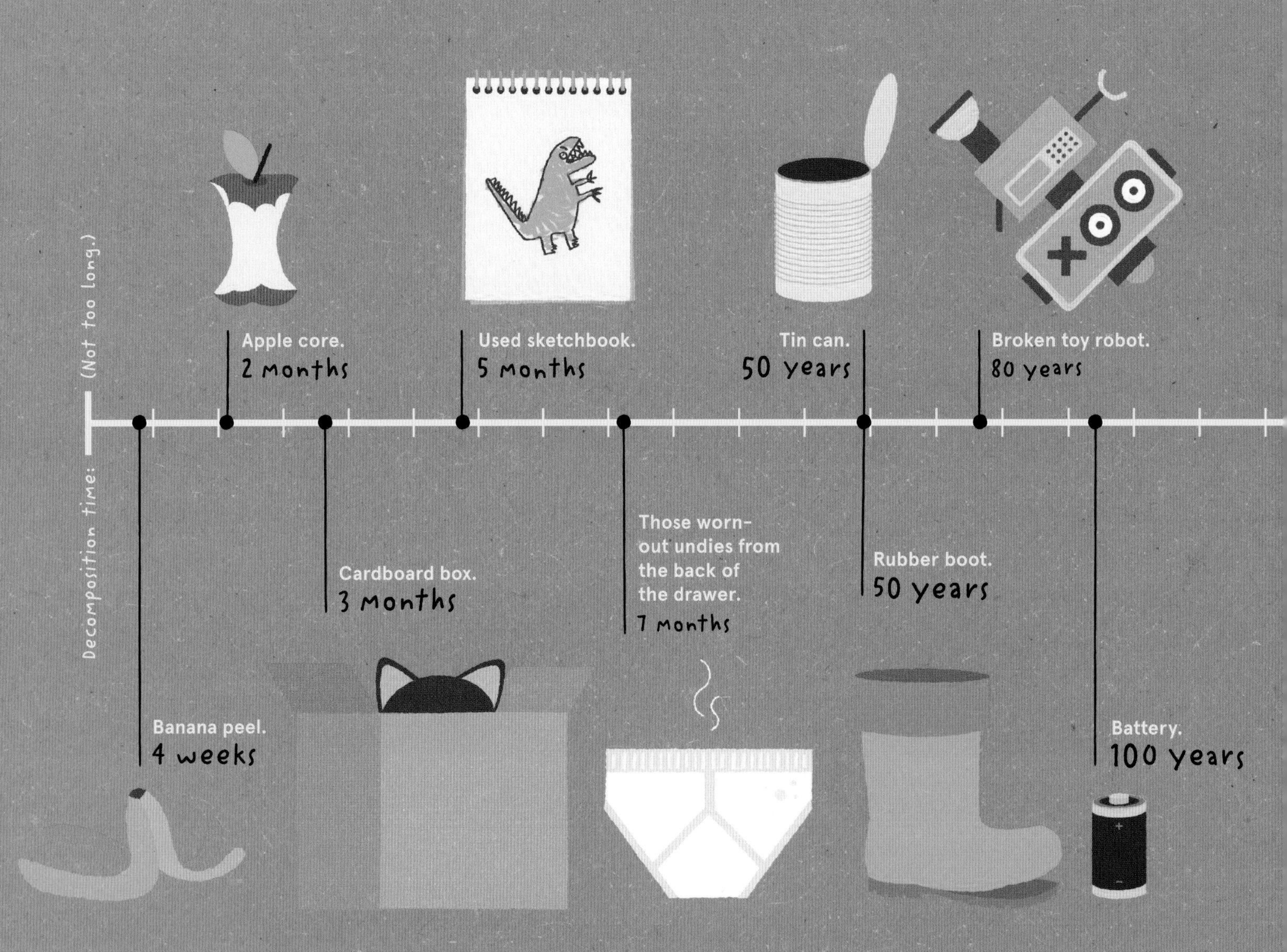

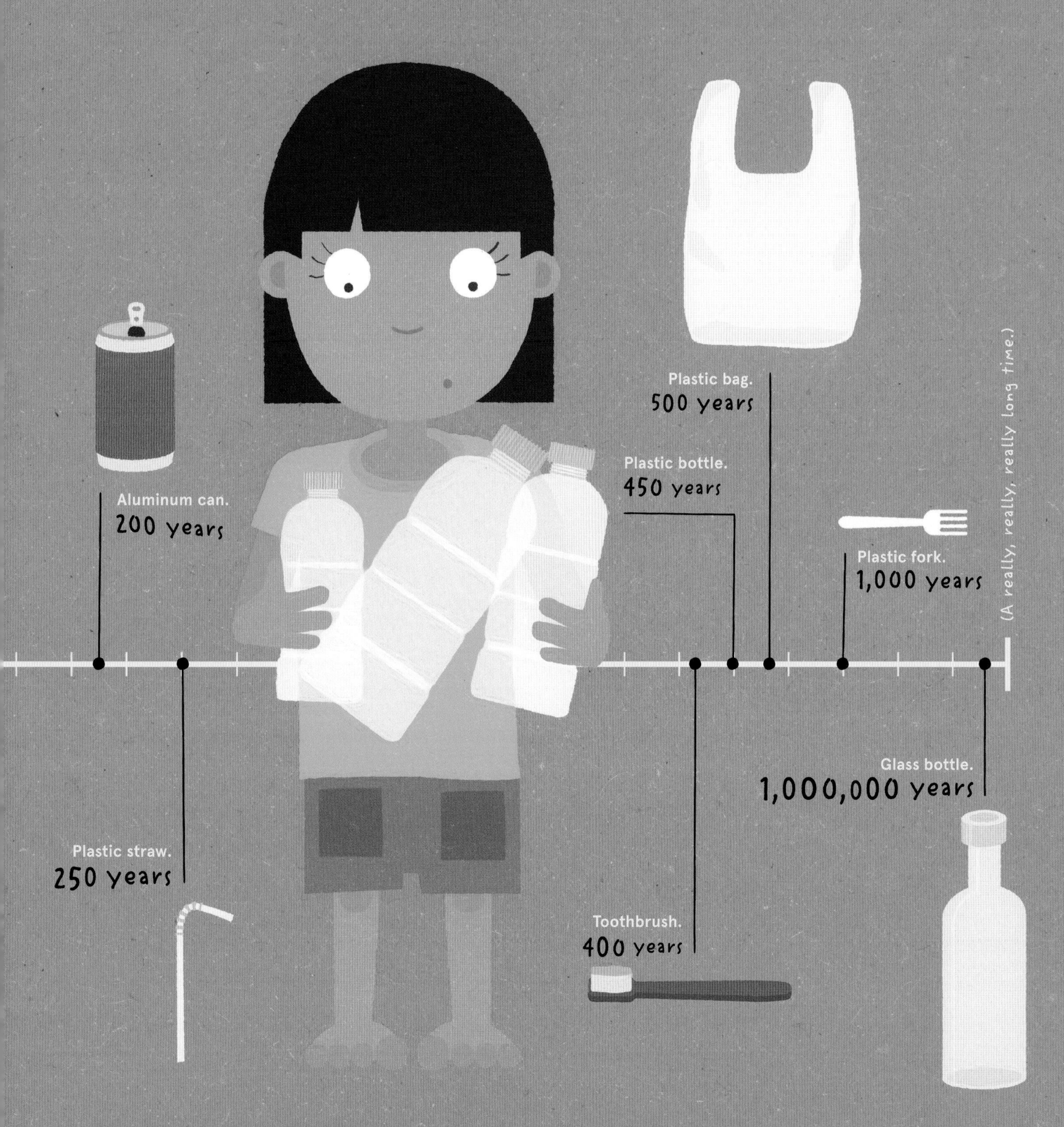

Did you see that glass bottle? A million years to decompose! That's why it is so important for us to deal with our waste in the very best way possible.

Our waste can hang around for generations to come, so it's up to all of us to make sure it ends up in the right place. But where does our household trash go once we're done with it?

Junk journey 1:

Landfill.

Most of the stuff we throw out at home is sent to a landfill. Landfills consist of ginormous stinky pits designed to bury and very, very slowly decompose our waste.

As their funky filling breaks down, landfill pits let off nasty greenhouse gases such as carbon dioxide (CO_2) and methane (CH_4), which are warming our planet too quickly. So the less waste we can send to landfills, the better.

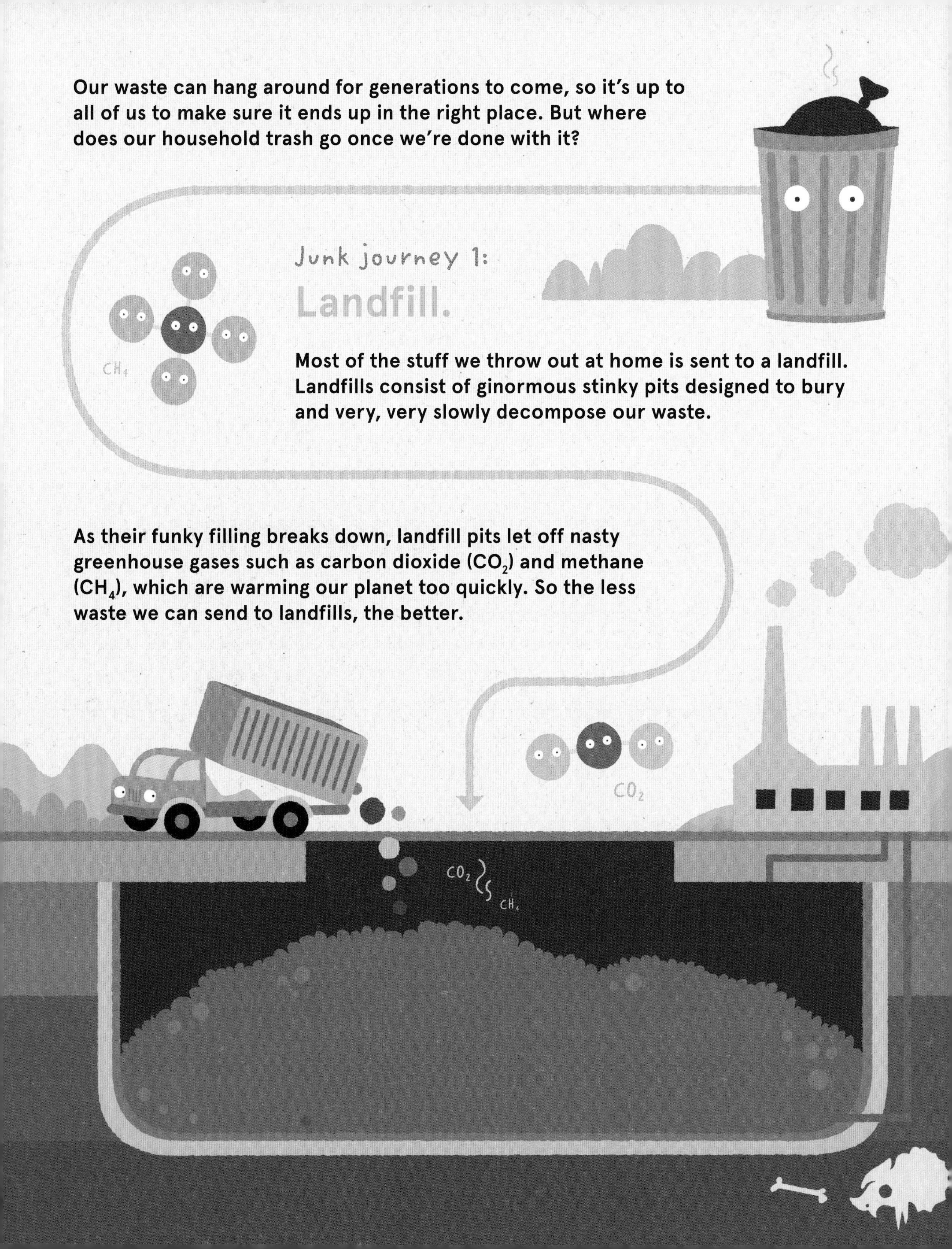

Junk journey 2:

Recycling.

Happily, we can avoid sending lots of our junk to landfills by recycling it!

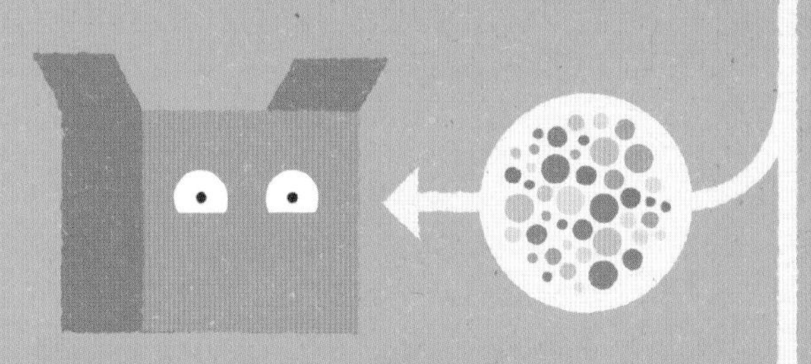

Rather than burying our trash, recycling means that a lot of our household waste can be cleaned, chopped up into its raw materials, and used again to make new recycled products.

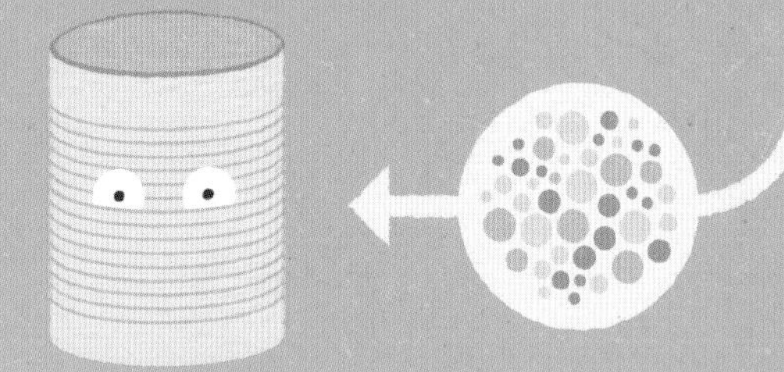

Making things from recycled materials is far less wasteful than making them from new materials.

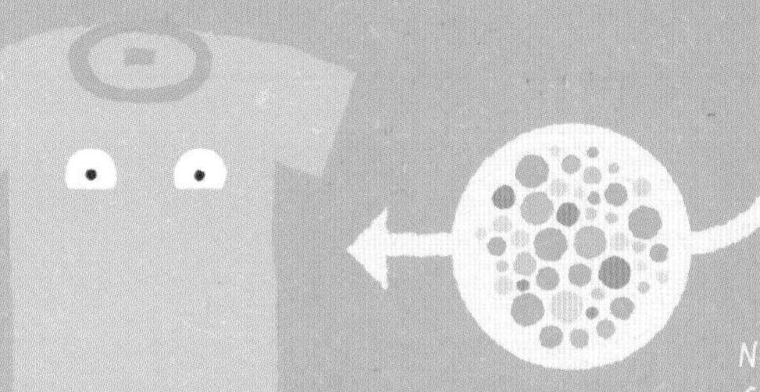

New clothing can be made from recycled plastic!

Junk journey 3:

Environment.

The most terrible place our garbage can end up is in the natural environment.

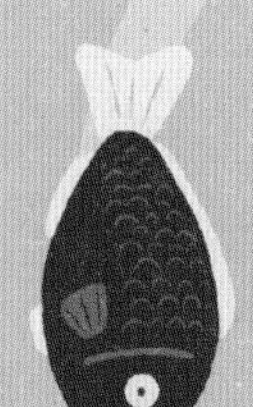

This happens when we are really careless with our trash and don't put it in a waste or recycling bin. Sadly, it is still very common for waste to enter our environment.

Most of this trash makes its way down drains to streams and rivers then eventually out to the ocean, where it has a devastating effect on marine creatures.

So when you see litter out in the wild, or even on the street, pick it up—if it is safe to do so—and put it in a trash can.

Food waste can be turned into plant food (making lots of tiny creatures very happy in the process).

Junk journey 4:

A new purpose.

Sometimes we can find new uses for things we're done with, or we can pass them on so that someone else can use them.

Clothes you've outgrown can be given to thrift stores or to younger kids you know.

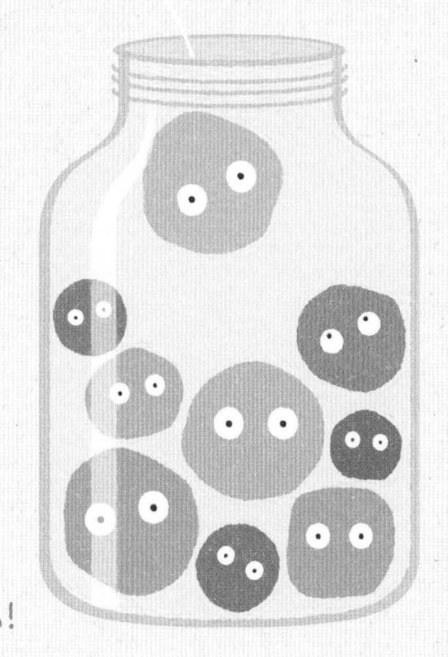

Old jars can be reused to house your rock collection!

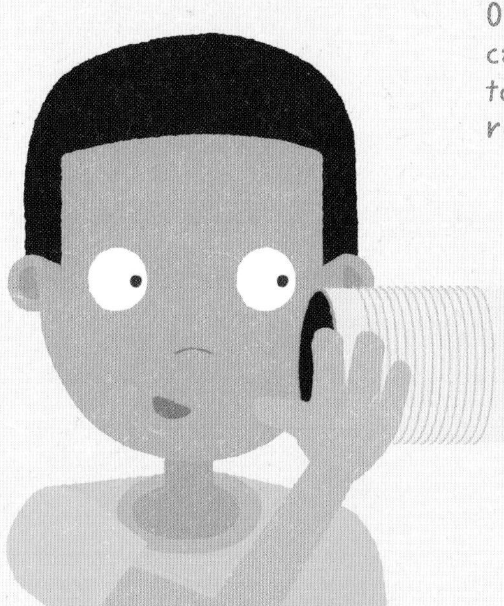

Tin cans can be turned into a telephone!

How can you help wallop waste?

As individuals we're only small, but there are nearly eight billion of us here on Earth. If each one of us does a little bit, we can make a huge impact. Here are five ways you can help your planet (while using more *R*s than Captain Hook) . . .

Renew.

Really
get involved!

1 Reduce!

Buy less. Buy better.

One of the best ways to reduce our waste is to buy less stuff! Lots of things that are made aren't really needed, or aren't made to last. So before you buy or use something, think about whether you really need it. If you really do need something, choose quality things over junky items that will break quickly.

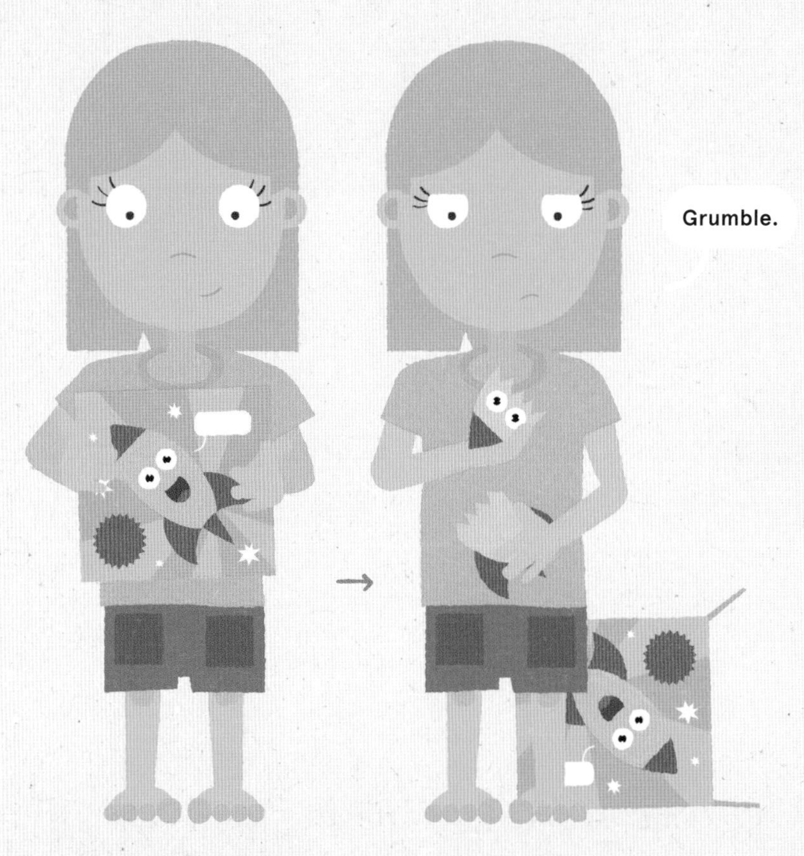

Fig 1. Cheap new toy purchased.
Fig 2. Less than a day later . . .

Unplug.

The electricity that powers most homes comes from wasteful and environmentally damaging fossil fuels like oil, gas, and coal. So while you might not see the changes directly, reducing the amount of energy you use at home will reduce your waste and help our precious planet to cool down. Here are five easy ways to help:

Switch off.
Turn off the TV or computer when nobody's using them. Turn off light switches.

Be cool.
Open windows and use fans rather than running the air conditioner in the summer.

Wrap up.
Put on hats and sweaters rather than turning up the heat during winter months.

No drips.
Turn off dripping taps and take short showers. It takes energy to clean and heat our water.

Be free.
Play outside when you can—read books, tell jokes! The very best things in life don't need electricity.

Booooooooo!

Single-use plastics.

One of the biggest waste challenges we face is plastic pollution in our oceans. This is caused largely by single-use plastic products—things that are made to be used only once and then thrown away.

Marine animals regularly mistake our floating plastic waste for food, with catastrophic consequences.

A floating plastic bag can look a lot like a delicious jellyfish to a hungry sea turtle.

You can help your planet by reducing your use of single-use plastics like disposable water bottles, food containers, and plastic wrap wherever you can. Encourage your family and friends to do the same. If we stop using these things, eventually companies will stop making them. If we don't buy junk, they won't make junk!

2 Reuse!

Reusing means that we keep stuff in use for as long as possible. Keep your eyes peeled for things that you think could be reused rather than sent to a landfill.

Try making puppets with worn-out socks! (Hint: wash the socks first—very few people enjoy a pungent puppetry performance.)

If it's broken, fix it!

Lots of things eventually break, but most things can be repaired. Often something that has been repaired will be more special to you—the repair will become a part of its story.

Ask a kind adult in your life to help you repair things.

Give it away.

If you can't reuse something but it is still in good condition, consider donating it to a thrift store. Thrift stores help our planet by preventing stuff from going to the landfill, as well as lessening the demand for new products.

Thrift stores include charity shops, junk shops, and more.

Here's how to reuse an old egg carton to grow your own vegetables!

You will need:

Used egg carton (no eggs required).

Gloves.

Vegetable seeds.

Teaspoon.

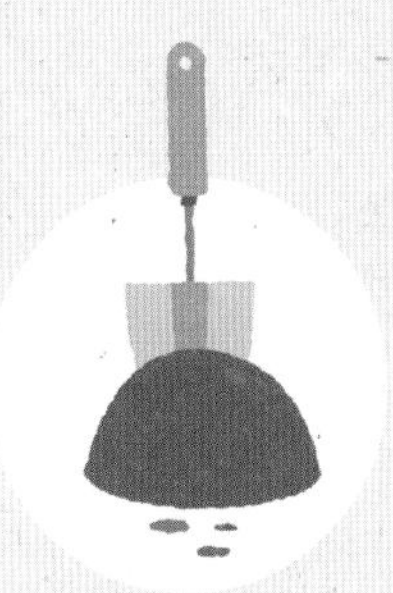
Potting mix (soil).

Water.

And here's how to do it:

1 Add soil.
Add potting mix to each section of the egg carton. Fill each part to half an inch below the top.

2 Plant seeds.
Plant two or three seeds per container and cover them with another layer of soil.

3 Tuck them in.
With your gloves on, gently press down on the top of the soil to help settle the seeds in.

4 Add water.
Add a few teaspoons of water to the soil, until the top layer looks nice and moist.

5 Find a nice spot.
Find your seeds a home in a nice bright place, being careful to avoid direct sunlight. Windowsills, balconies, or porches make lovely homes for young plants.

6 Take care.
Check your seeds each morning and each night before bed. If the top layer looks a bit dry, add a few more teaspoons of water. The soil should be moist, but not wet.

7 Grow on!
You should see green shoots in about a week. When they're big enough, plant your seedlings directly into the ground or in a bigger pot, egg carton and all!

↓

The **roots** will grow through the egg carton and the cardboard will turn into **compost**!

3 Recycle!

Recycling our household waste reduces the amount of trash we send to the landfill. It is one of the easiest ways to help our planet.

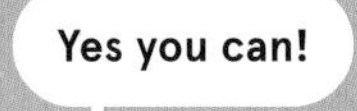

Because the recycling process reuses materials that have already been made, it also reduces the amount of resources we need to take from the planet.

When we recycle something—rather than sending it to the landfill, where it will never be seen again—we are giving it a whole new glorious life!

What can you recycle at home?

Unfortunately, not all household junk can be recycled. Here are a few general rules on what should (and should not) be put into your home recycling bin:

Yes!

Hard plastic.

- Soda and juice bottles
- Shampoo bottles
- Fruit containers
- Takeout food containers

Metal.

- Aluminum cans
- Tin cans
- Aerosol cans
- Foil

(If scrunched into a ball.)

Glass.

- Food jars
- Green bottles
- Brown bottles
- Clear bottles

Paper.

- Sketchbooks
- Office paper
- Magazines/newspapers
- Junk mail
- Cardboard boxes

Please check cardboard boxes for pets before recycling.

Nope.

Household stuff.

- Old clothes and fabric
- Broken toys
- Ceramics and ovenware
- Polystyrene packaging
- Broken light bulbs

Medical waste.

- Bandages
- Diapers
- Baby wipes
- Syringes
- Glass eyes

Soft plastic.

- General soft packaging (such as bread bags, frozen food bags, chocolate bar wrappers, and cereal box liners)
- Bubble wrap
- Plastic bags

In many places, you can bundle these up and take them to dedicated soft-plastic recycling centers. Woohoo!

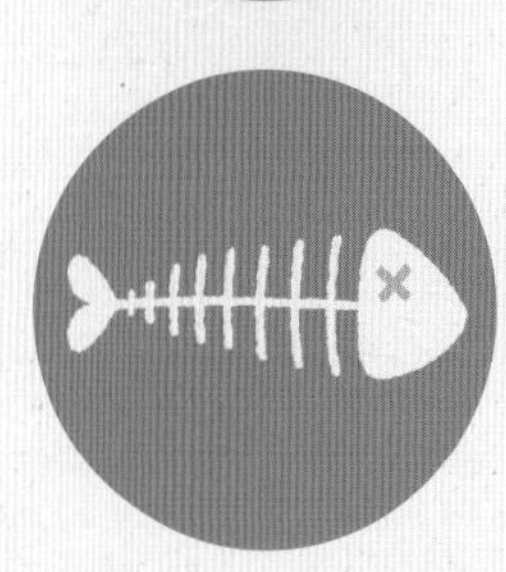

Organic waste.

- Food waste
- Garden waste
- Hair clippings
- Pet poop
- Ex-pets

4 Renew!

A major waste problem is food waste, which accounts for almost half of all waste in landfills in developed countries.

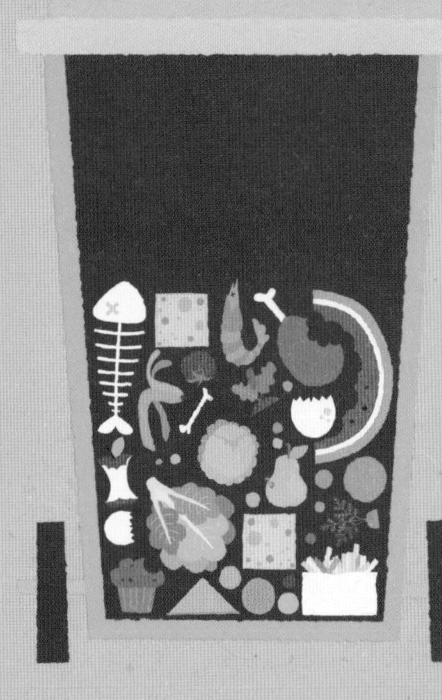

That's around **half** of your **household trash** each week!

It is estimated that we throw out around one in every five bags of food we bring home.

When it finally breaks down in the landfill, food waste produces methane. This is a natural gas, but unfortunately methane traps heat in our atmosphere, which makes our planet hotter and causes all kinds of problems.

Incredibly serious footnote: As we have all come to understand from dealing with the flatulence of our fellow humans, not all naturally occurring gases are entirely pleasant.

But here's the good news! Food waste is organic, meaning that it grew in—or on—the earth. So rather than chucking it into the landfill to create harmful gases, a lot of our food waste can be put back into the natural cycle and renewed.

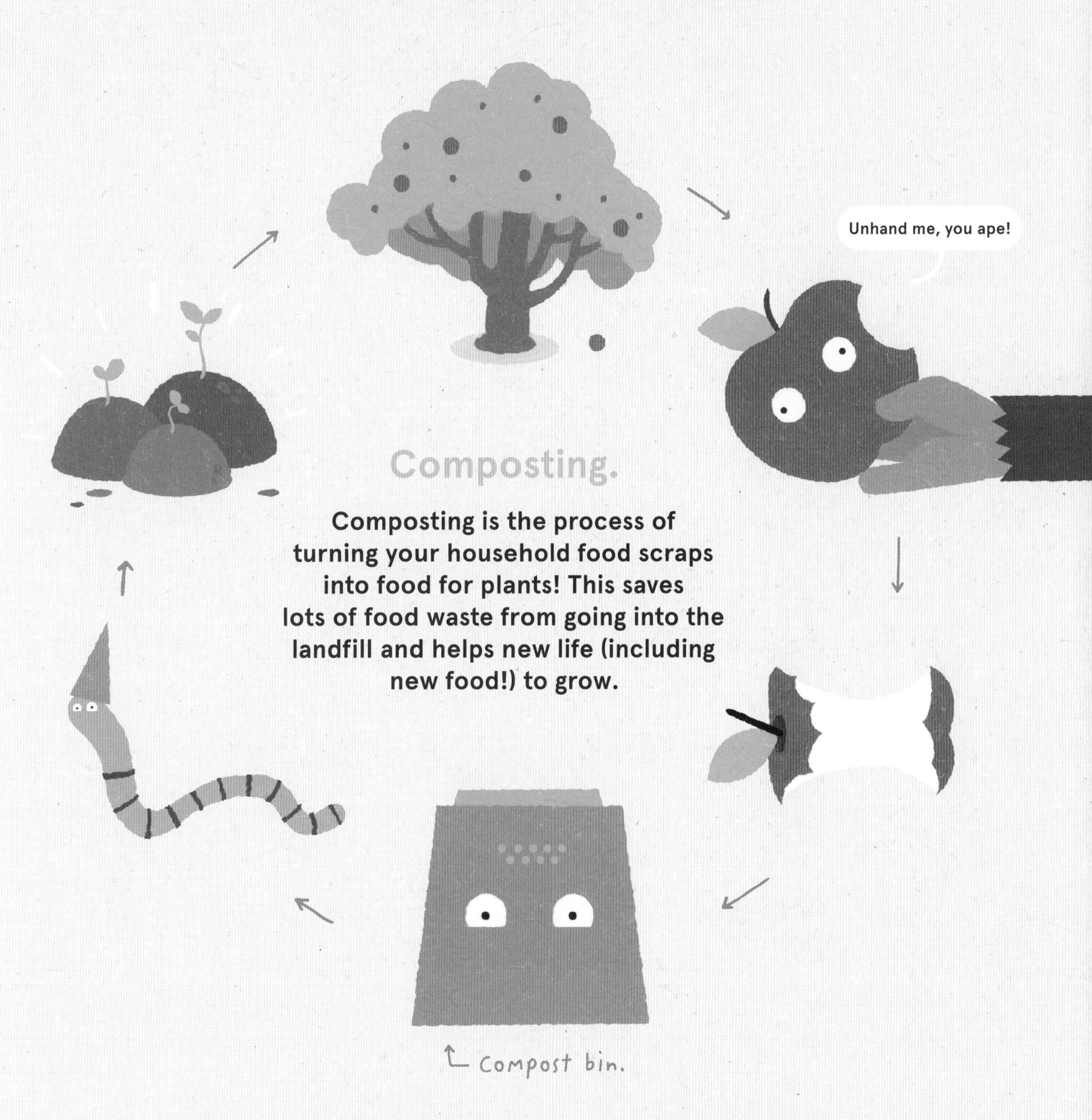

Composting.

Composting is the process of turning your household food scraps into food for plants! This saves lots of food waste from going into the landfill and helps new life (including new food!) to grow.

Through composting, you can reduce your food waste and help make your little corner of the earth a richer place. Let's find out how . . .

Did somebody say compost?

How to compost.

There are many ways to turn your food scraps into compost. Whether you live in an apartment in the city or on a farm out in the country, you can turn your food scraps into plant food!

This method is designed for a small space in a backyard. If you don't have a backyard, there are heaps of composting alternatives out there for apartments or small outdoor spaces.

You will need:

A sunny spot in a garden or backyard.

Compost bin.

A pile of food scraps.

Dry brown leaves, twigs, grass clippings, or other dry (dead) organic stuff.

Water.

What can be composted?

Vegetables.

Fruit.

Grains.

Tea and coffee.

Garden waste.

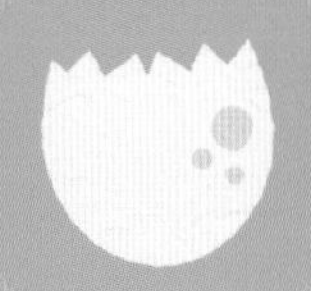
Eggshells.

Nope.

Meat.

Cheese.

Fats.

Poop.

Non-organics.

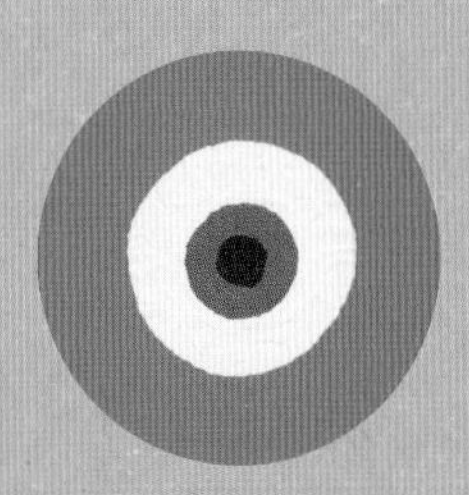
Glass eyes.

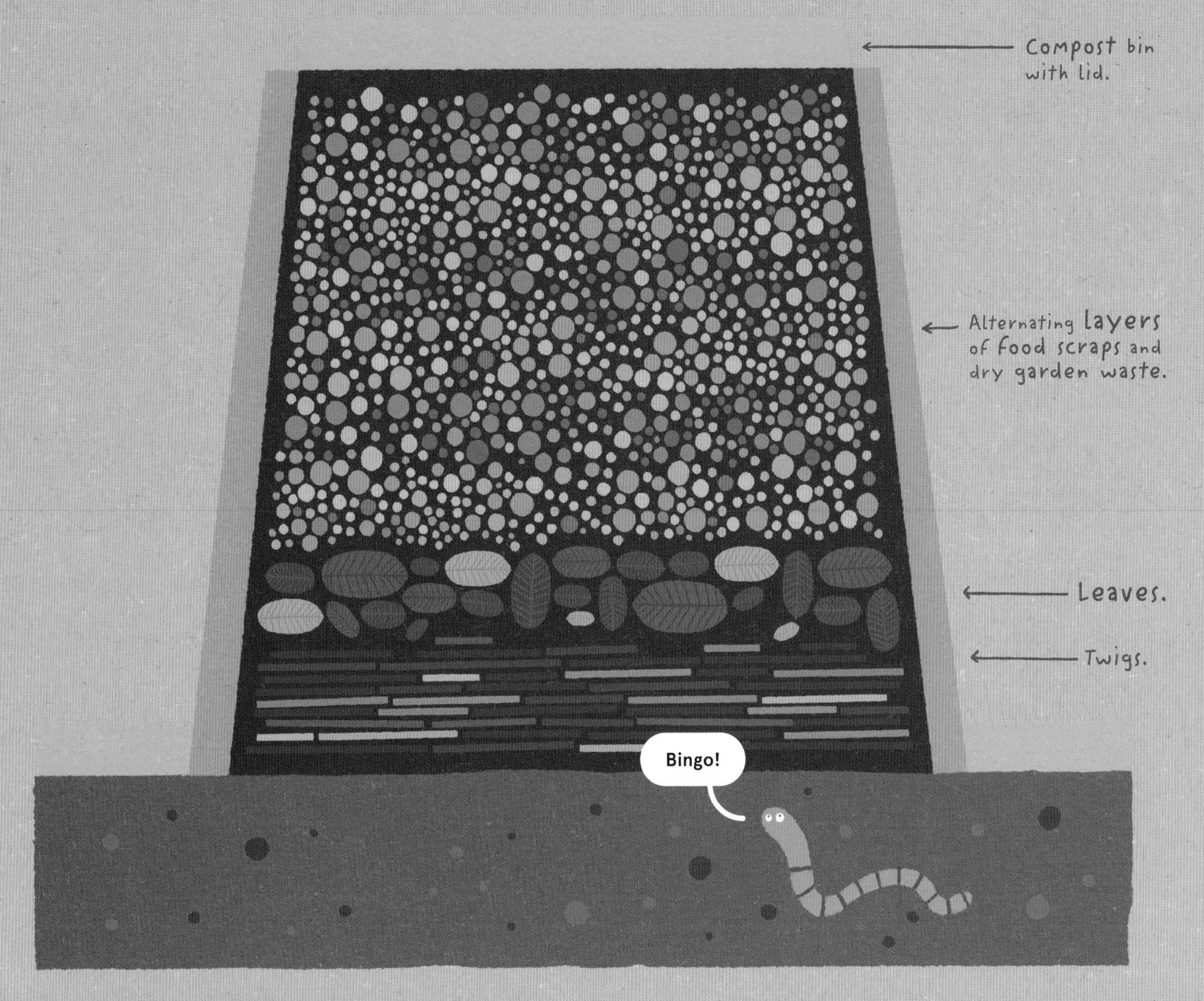

1 Find your spot.

Place your compost bin in a sunny backyard spot that's easy to get to each day, but not in the way. The bin should be placed on bare soil so that the worms (and company) can reach the food scraps.

2 Create a base layer.

Add a layer of dry twigs, sticks and leaves, covering the soil at the bottom of the bin. This will help keep air flowing around your compost, which helps the decomposers do their thing.

3 Add food scraps.

Add your first layer of food scraps on top of the dry twigs and leaves. Keep adding your scraps to the bin each day until you have a layer of food scraps about 6 inches deep.

4 Add a dry layer.

Cover the first layer of food scraps with garden waste—brown, dead organic stuff such as leaves, grass clippings, small twigs, and the like. Try to keep your layer of dry garden waste around 6 inches deep.

5 Repeat!

Repeat with alternating layers of food scraps and dry garden waste until your compost bin is full. Remember that you are saving all of these food scraps from being buried in landfills!

6 And now, we wait.

It can take anywhere from 3 to 12 months to produce good compost, depending on your local climate, what's gone into the compost, and how keen the worms are. So. Prepare. To. Be. Patient.

7 Wait some more.

3 to 12 months is a really, really, really long time! While you're waiting, begin to think about where you might use your compost when it's ready.

8 Are we done?

Finished compost is dark and crumbly in texture, with a nice earthy smell. Ask a friendly adult to help you figure out when it's good to go.

9 Feed some plants!

When it's ready, you can use your compost to feed house plants, your garden, and local trees (you could even plant a banana tree if you're feeling very brave).

5

Really get involved!

When you feel strongly about a particular issue, take action to make your thoughts heard. You can help make a positive change in the way others think.

← You might choose to join a march in support of a cause you believe in.

Or you might write a letter to your local supermarket asking them to kindly use less plastic packaging on their fruit and veggies.

↑ (This is bananas.)

Speak up.

Talk to your family and friends about reducing, reusing, and recycling waste. The more people who know about this stuff, the more of a difference we can make for our planet.

Clean up.

Many good folks and families choose to have fun together while removing litter from their little corner of the earth. Think about joining a community clean-up at your local park, playground, or beach.

Step up.

Actively support organizations and people who are trying their best to help our planet. Try to avoid those who act only in their own interest, and not in the best interest of the earth.

You, me, your family, your friends (yep, even the wild ones)—we are the temporary custodians of this beautiful little planet called Earth.

This is our only home, and it is up to all of us to look after it. Use the ideas from this book to help clean up our precious planet, and in time more people will join you.

Positive change comes when good people do good things.
You can be the change.
Your planet needs you!

FOR ARTHUR
WITH LOVE XX

This book is a very brief introduction to the complex concept of waste. It is intended as an optimistic first step toward positively engaging even the youngest minds with a subject that is ultimately critical to the life of every living thing on this planet. I hope the ideas in these pages are accessible enough for those youngest minds to grasp, and simple enough for slightly older minds to begin to put into daily practice.

PHILIP BUNTING

BLOOMSBURY CHILDREN'S BOOKS
Bloomsbury Publishing Inc., part of Bloomsbury Publishing Plc
1385 Broadway, New York, NY 10018

BLOOMSBURY, BLOOMSBURY CHILDREN'S BOOKS, and the Diana logo
are trademarks of Bloomsbury Publishing Plc

First published in Australia in 2020 by Little Hare Books, an imprint of Hardie Grant Egmont, Richmond, Victoria
Published in the United States of America in February 2022
by Bloomsbury Children's Books

Text and illustrations copyright © 2020 by Philip Bunting

All rights reserved. No part of this publication may be reproduced or transmitted in any form or by any means, electronic or mechanical, including photocopying, recording, or any information storage or retrieval system, without prior permission in writing from the publisher.

Bloomsbury books may be purchased for business or promotional use.
For information on bulk purchases please contact
Macmillan Corporate and Premium Sales Department at
specialmarkets@macmillan.com

Library of Congress Cataloging-in-Publication Data
available upon request
ISBN 978-1-5476-0792-1 (hardcover) • ISBN 978-1-5476-0801-0 (e-book)
ISBN 978-1-5476-0796-9 (e-pdf)

Typeset in Apercu
Printed in China by Leo Paper Products, Heshan, Guangdong
10 9 8 7 6 5 4 3 2 1

To find out more about our authors and books visit www.bloomsbury.com
and sign up for our newsletters.

Acknowledgment of Country: I would like to acknowledge the traditional custodians of the land on which I live and work, and I pay respect to the Gubbi Gubbi nation. I pay respects to the Elders of the community and extend my recognition to their descendants. —Philip Bunting.

Tomato seeds.
Tomato seeds.